What is Relativity?

L. D. Landau & G. B. Rumer
(Translated by N. Kemmer)

DOVER PUBLICATIONS, INC.
MINEOLA, NEW YORK

Bibliographical Note

This Dover edition, first published in 2003, is an unabridged reprint of a translation of the Russian book *What is the Theory of Relativity* (Moscow, 1959), published by Basic Books, Inc., New York, in 1960.

Library of Congress Cataloging-in-Publication Data

Landau, L. D. (Lev Davidovich), 1908-1968.
   [Chto takoe teoriia otnositel'nosti. English]
     p. cm.
   Includes index.
   Originally published: New York : Basic Books, 1960.
   ISBN 0-486-42806-0 (pbk.)
    1. Relativity (Physics) I. Rumer, IU. B. (IUrii Borisovich) II. Title.

QC6 .L26213 2003
530.11—dc21

2002035112

Manufactured in the United States of America
Dover Publications, Inc., 31 East 2nd Street, Mineola, N.Y. 11501

Contents

What is Relativity?

1

Familiar Examples of Relativity

Does Every Statement Have a Meaning?

Evidently not. Even if you choose perfectly sensible words and put them together according to all the rules of grammar, you may still get complete nonsense. For instance the statement "This water is triangular" can hardly be given any meaning.

Unfortunately, however, not all examples of nonsense are so obvious and it often happens that a statement appears perfectly sensible at first sight but proves to be absurd on closer examination.

Right and Left

Look at the drawing on the next page. On which side of the road is the house—on the right or on the left? It is impossible to answer this question directly.

If you are walking from the bridge to the wood, the house will be on the left-hand side, but if you go from the wood to the bridge, you find the house on the right. Clearly, to speak of the right- or left-hand side of the road you must take into account the direction relative to which right or left is indicated.

It does make sense to speak of the right bank of a river, but only because the current determines the direction of the river. Likewise we can only say that cars keep to the right because the movement of a car singles out one of the two possible directions along the road.

[3]

We see that the notions "right" and "left" are relative: They acquire meaning only after the direction relative to which they are defined has been indicated.

Is It Day or Night Just Now?

The answer depends on where the question is being asked. When it is daytime in Moscow it is night in Vladivostok. There is no contradiction in this. The simple fact is that day and night are relative notions and our question cannot be answered without indicating the point on the globe relative to which the question is being asked.

Who Is Bigger?

In the first drawing on the opposite page, the shepherd is obviously bigger than the cow; in the second the cow is bigger than the shepherd. Again there is no contradic-

tion. The reason is that the two drawings have been made by people observing from different points: One of them stood closer to the cow, the other closer to the shepherd. The picture is determined not by the actual sizes of the objects but by the angles at which they are seen. Evidently such angular dimensions of objects are relative. It makes no sense to speak of the angular dimensions of objects without indicating the point in space from which they are observed. For instance, to say, "This tower is seen

under an angle of 45°," is to say precisely nothing. But the statement that the tower is seen under an angle of 45° from a point 15 meters (about 50 feet) away has a definite meaning; from this you can conclude that the tower is 15 meters high.

The Relative Seems Absolute

If the point of observation is moved a small distance, angular dimensions also change only by a small amount. That is why angular distances are often used in astronomy. A star map usually gives the angular distance between stars, *i.e.*, the angle at which the distance between the two stars is observed from the surface of the earth.

We know that however much we move about on the Earth and whatever points on the globe we choose to observe from, we always see the stars in the sky at the same distances from each other. This is because the stars are such unimaginably large distances away from us that, in comparison, our movements on the Earth are negligible and can safely be forgotten. In this particular case we can therefore use angular distances as absolute measures of distance.

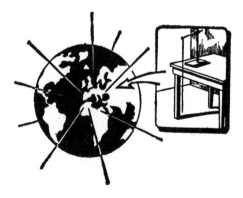

If we make use of the Earth's motion around the sun, it becomes possible to observe changes in angular distances between stars even though these changes are very small. But if we were to move our point of observation to some other star, such as Sirius, all angular distances would change so much that stars far apart in our sky might then be close together, and vice versa.

The Absolute Proves To Be Relative

We often use the words "up" and "down." Are these notions absolute or relative?

At different times in history the answers given to this question have differed. As long as people didn't know that the Earth is a sphere and thought that it was flat, like a pancake, the vertical was taken to be an absolute direction. It was taken for granted that this vertical direction was the same at all points on the Earth's surface, making it perfectly natural to speak of an absolute "up" and an absolute "down."

When it was proved that the Earth is a sphere the vertical began to totter—in people's minds. For if the Earth is a sphere, the direction of the vertical depends decisively on where on the Earth's surface the vertical is drawn.

Different points on the Earth will have different verticals. The notions "up" and "down" now cease to have meaning unless the point on the Earth's surface to which they refer is defined. Thus these notions change from absolute to relative ones. There is no unique vertical direction in the universe. Therefore, given any direction in space, we can find a point on the Earth's surface at which this direction is the vertical.

"Common Sense" Tries To Protest

Nowadays all this seems obvious and indisputable to us. But the record of history shows that in the past it was not so easy for mankind to understand the relativity of up and down. People have the tendency to ascribe absolute meaning to notions as long as their relative nature is not evident from everyday experience (as in the case of "right" and "left").

Remember the ridiculous objection to a spherical Earth which has come down to us from the Middle Ages: How could people possibly walk about upside-down?

The flaw in the reasoning here is that of not recognizing that, since the Earth is a sphere, the vertical is relative.

If you refuse to accept the principle of relativity for the vertical direction and assume, say, that the direction of the vertical in Moscow is absolute, you are bound to admit that the inhabitants of New Zealand walk upside-down. But if you do this, you should remember that for New Zealanders, the people in Moscow are the ones walking upside-down. This is no contradiction, because the notion of the vertical is not absolute but relative.

We should make a note of the fact that we begin to sense the real significance of the relativity of the vertical only when we consider two sufficiently distant parts of the Earth's surface, such as Moscow (or New York) and New Zealand. If we are concerned with two neighboring spots such as two houses in Moscow, we can for practical purposes take all verticals to be parallel, which means taking the vertical direction there to be absolute.

Only when we are obliged to deal with regions of a size comparable with the whole surface of the Earth do we find that trying to make use of an absolute vertical leads to absurdities and contradictions.

Our examples have shown that many of the notions in everyday use are relative, which means that they acquire meaning only when the conditions of observation are stated.

2

Space Is Relative

The Same Place or Not?

We often say that this event and that happened at the same place, and we are so used to saying this sort of thing that we tend to ascribe absolute meaning to such a statement. In truth it has no meaning at all! It is no better than saying "It is now five o'clock," without indicating whether it is supposed to be five o'clock in Moscow or in Chicago.

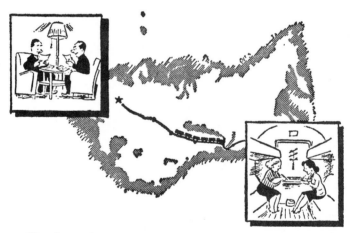

To show that this is so, let us imagine that two lady passengers traveling in the express from Moscow to Vladivostok have agreed to meet every day during the journey at the same place in the train, in order to write letters to their husbands. Their husbands will hardly agree that their wives have been meeting each day at one and the same place in space. On the contrary they will have every reason to state that the places at which their wives met from day to day were hundreds of miles apart. They received letters from Yaroslavl and from Perm, from Sverdlovsk and from Tyumen, from Omsk and from Khabarovsk.

So the two events—writing letters on the first and the second day of the journey—occurred at one and the same place from the point of view of the traveling ladies, but were many hundreds of miles apart from the point of view of their husbands.

Who is right, the travelers or their husbands? We have no reason to give preference to either. Quite clearly the notion "at the same place in space" has only relative meaning.

In the same way, the statement that two stars in the sky coincide has meaning only if we indicate that the observation is made from the Earth. We can speak of two events coinciding in space only if we indicate some objects relative to which the position of the events is determined.

Thus the notion of position in space is again a relative one. When we speak of the position of an object in space, we always assume this to mean its position relative to other objects. We have to admit that it is meaningless to demand that the whereabouts of one object be fixed without reference to other objects.

How Does a Body Move in Reality?

It follows from all this that the notion of "displacement of a body in space" also is relative. If we say that an object was displaced, this means no more than that it changed its position relative to other objects.

If the movement of an object is observed from different laboratories which are in motion relative to each other, the movement of the body will appear quite different.

A stone is dropped from a flying plane. Relative to the plane, the stone falls in a straight line; relative to the earth it describes a curve called a parabola.

But how does the stone move in reality?

This question has as little meaning as the question: "At what angle does one see the moon in reality?" Do you mean the angle at which it is seen from the sun or from the Earth?

The geometric form of a curve described by an object in motion is of the same relative nature as the photograph of a building. Depending on whether a house is photographed from the front or the back, different views are obtained. In the same way, depending on whether we observe the movement of an object from one laboratory or from another, we obtain different shapes of curve to describe its motion.

Are All Points of View Equivalent?

If in observing the motion of an object in space we were only interested in studying the shape of its trajectory (as the curve along which it moves is called), we would approach the question of choosing our place of observation with an eye to the convenience and simplicity of the resulting picture.

A good photographer choosing a place for his exposure is concerned with composition as well as with the beauty of the intended photograph.

However, when we study the displacement of objects in space we are interested in rather more. We do not want to know just the trajectory. We also want to be able to predict the trajectory along which the object will move under given conditions. In other words, we want to know the laws governing motion, the laws that force a body to move in just this way and no other.

Let us examine the question of the relativity of motion from this point of view; we shall see that not all situations in space are equivalent.

If you go to a photographer for a passport photograph, you naturally want him to take a photograph of your face, not of the back of your head. This demand determines

the point in space from which the photographer must take the picture. We would say that any other position does not satisfy the condition imposed.

Rest Is Found!

The motion of objects is influenced by external actions. We call these actions forces. By studying the result of these actions we find that a completely new approach to the question of motion becomes possible.

Let us assume that we have available an object which is not acted on by any forces. We may position ourselves in different ways to observe it and accordingly will see it moving in different, more or less extraordinary ways. But it is impossible to deny that the most natural position for the observer will be the one from which the body appears to be simply at rest.

In this way we can give a completely new definition of rest, which does not depend on the displacement of the object in question relative to other objects. This is how we do it: An object on which no external forces are acting is in a state of rest.

The Laboratory at Rest

How can we bring this state of rest into being? When can we be sure that no forces whatsoever are acting on an object? Evidently we should remove the object as far as possible from all other objects that might act on it.

From such objects at rest we could at least in imagination construct a whole laboratory, and we could then speak of the properties of motion observed from this laboratory, which we call a laboratory at rest.

If the properties of a motion observed from any other laboratory differ from the properties of motion in the laboratory at rest, we have every right to assert that the first laboratory is moving.

Is the Train Moving?

Having established that in moving laboratories motion takes place according to other laws than in laboratories at rest, it would seem that the concept of motion loses its relative character. When henceforth we talk of motion, we may simply take this to mean motion relative to rest, and call such motion absolute.

But will any displacement of a laboratory lead us to observe that the laws of motion in it differ from those in a laboratory at rest?

We are sitting in a train running at constant speed along a straight track. We start observing the motion of objects in the railway carriage and comparing them with what takes place in a train at rest.

Everyday experience tells us that in such a train, moving in a straight line and at uniform speed, we do

not notice any differences, compared to the motions of objects observed in a stationary train. If a ball is thrown up vertically inside a railway carriage, the ball will fall back into one's hands. A falling object will not describe a curve similar to the one drawn on the next page.

Apart from the shakes and bumps that are inevitable in practical running conditions, things take place in a uniformly moving railway carriage exactly as in a carriage at rest.

It is different if the carriage accelerates or slows down. In the former case we experience a jerk backward, in the latter, forward; in both we distinctly notice the difference compared to rest.

If the railway carriage continues to move steadily but with changing direction, we shall also sense it: At sharp right-hand curves we get thrown to the left; at left-hand turns, to the right.

Generalizing these observations, we reach this con-

clusion: As long as a laboratory moves at uniform speed and in a straight line relative to a laboratory at rest, it is impossible to discover in it any differences in the behavior of objects compared to the laboratory at rest. But as soon as the speed of the moving laboratory changes in magnitude (acceleration or deceleration) or in direction (curve), this immediately shows up in the behavior of objects in it.

Rest Is Lost For Good

The uniform straight-line motion of a laboratory has a surprising property. It does not influence the behavior of objects within it. This forces us to re-examine the notion of rest. It turns out that the state of rest and the state of uniform straight-line motion do not differ from each other in any way. A laboratory moving uniformly in a straight line relative to a laboratory at rest can itself be taken to be a laboratory at rest. This means that there exist not one absolute state of rest but an innumerable multitude of different "rests." There exist not one laboratory "at rest" but an innumerable multitude of laboratories "at rest," all moving in straight lines and uniformly at different speeds relative to one another.

Since rest has proved to be not absolute but relative, we see that motion also must be relative in all cases.

We must always answer the question, "Relative to which state of 'rest' is the motion being observed?"

So we have arrived at a most important law of nature, which is usually called the principle of the relativity of motion. It says: In all laboratories moving in straight lines and uniformly with respect to each other the motion of objects takes place according to the same laws.

The Law of Inertia

It follows from the principle of relativity of motion that an object on which no external force acts can be not only in a state of rest but also in a state of uniform straight-line motion. This statement is known is physics as the law of inertia.

However, in everyday life this law is, so to speak, obscured and not directly in evidence. According to it, an object in a state of uniform straight-line motion should continue this motion indefinitely, unless external forces act on it. But we know from observation that objects to which we do not apply forces tend to come to rest.

The explanation is that all objects that we are able to observe are subject to certain external forces—the forces of friction. The condition required for the law of inertia to hold good does not exist—namely, the total absence of external forces acting on the body. But if we keep improving the conditions of the experiment and reducing the frictional forces, we can get closer and closer to the ideal condition necessary. In this way we can prove that this law is also valid for motions observed in everyday life.

The discovery of the relativity principle for motion is one of the greatest of all discoveries. The development of physics would have been quite impossible without it. We owe the discovery to the genius of Galileo Galilei, who took a brave stand against the teachings of Aristotle, which dominated men's minds in his day and were supported by the authority of the Catholic Church. Aristotle held that motion is possible only when forces act and immediately ceases without them. By a number of brilliant experiments Galileo showed that, on the contrary, it is the force of friction that causes moving objects to stop and that in the absence of this force a body once set in motion would continue to move forever.

Speeds Too Are Relative!

A result of the principle of relativity of motion is that there is just as little meaning in talking of uniform straight-line motion of a body with a certain speed, without indicating relative to what rest laboratory the speed is measured, as in speaking of geographical longitude without specifying in advance from which meridian it is to be reckoned.

Speed, too, proves to be a relative notion. If we determine the speed of one and the same body relative to different laboratories at rest, we shall obtain different results. But at the same time, any change of speed by acceleration or deceleration or by a change of direction has an absolute meaning and does not depend on what laboratory at rest is used for the observation.

3

The Tragedy of Light

Light Does Not Spread Instantaneously

We have convinced ourselves that the principle of relativity holds for motion and that there exist an innumerable multitude of laboratories "at rest." The laws of motion for material objects are the same for all these laboratories. Now, one form of motion appears at first sight to contradict the principle we have just established. This is the movement of light.

Light travels with the enormous speed of 300,000 kilometers (about 186,000 miles) per second, but nevertheless not instantaneously.

It is hard to imagine such an enormous speed, for the speeds we meet in everyday life are always immeasurably smaller. For instance, even the speed reached by a recent Soviet cosmic rocket is only 12 kilometers (7.5 miles) per second. Of all the objects with which we deal, the fastest moving is the Earth in its revolution around the sun. But its speed is still only 30 kilometers per second.

Can the Speed of Light Be Altered?

In itself, the enormous speed of light is not particularly surprising. The astonishing thing is that this speed has the property of always being strictly constant.

The motion of any object can always be artificially

slowed down or speeded up. Even a bullet. Put a box of sand in the path of a flying bullet. When it penetrates the box, it loses part of its speed and continues on its way more slowly.

For light, things are quite different. While the speed of a bullet depends on the kind of gun firing it and on the properties of the gunpowder, the speed of light in empty space is one and the same for all sources of light.

Put a glass plate in the way of a ray of light. While the ray is passing through the plate, its speed decreases, for the speed of light in glass is less than in empty space. But when the ray emerges from the plate, the light will again travel with a speed of 300,000 kilometers per second.

Quite unlike any other motion, the movement of light in empty space has the important property that it cannot be decelerated or accelerated. Whatever changes a ray of light may suffer inside a piece of matter, it continues to move with its original speed as soon as it emerges into empty space.

Light and Sound

In this respect the travel of light is more similar to the travel of sound than to the motion of ordinary objects. Sound is a vibrating motion of the substance in which it is traveling. Its speed is therefore determined by the nature of the substance and not by the properties of the instrument causing the sound. Like the speed of light, the speed of sound cannot be diminished or increased, even by making the sound pass through all kinds of material objects.

If, for instance, we put a metal barrier across its path,

sound will change its speed inside the barrier but will regain its initial speed as soon as it returns into the original substance.

Take a bell-jar, connect it to a pump, and put an electric bulb and an electric bell inside it. Then start pumping out the air. The sound of the bell will get weaker and weaker until it becomes quite inaudible, while the electric lamp goes on shining. This experiment shows directly that sound can spread only through a material medium, while light can travel in a vacuum. This is the essential difference between them.

The Principle of Relativity of Motion Seems To Be Shaken

The enormous but still not infinite speed of light in empty space comes into conflict with the principle of relativity of motion.

Imagine a train moving with the enormous speed of 240,000 kilometers per second. Assume we are at the front end of the train and that a lamp is switched on at the rear. Let us see what will happen if we measure the time taken by the light to pass from one end of the train to the other.

Seemingly this time will differ from the corresponding time measured in a stationary train. For relative to a train moving with a speed of 240,000 kilometers in a second the light should have a speed (in the direction of the train's forward motion) of only 300,000 − 240,000 = 60,000 kilometers per second. (It is as if the front end of the train runs away from the light.) If we put a lamp at the head of the train and measure the time taken by the light to get to the last carriage, we would expect the speed of light in the opposite direction to the train's motion to be 240,000 + 300,000 = 540,000 kilometers per second (because the light and the rear carriage move toward each other).

So we find that in the moving train, light should travel with different speeds in opposite directions, while in a stationary train its speed should be the same both ways.

For a bullet things are quite different. Whether we shoot the bullet in the direction of the train's motion or in the opposite direction, its speed relative to the walls of the carriage will always be the same—equal to its speed in a stationary train. This is because the speed of the bullet is influenced by the speed of the moving rifle. But as we said before, the speed of light does not change if the speed of the lamp changes.

Our arguments seem to show clearly that the behavior of light drastically contradicts the principle of relativity of motion. While a bullet has the same speed relative to the carriage walls in a train at rest and in a train in motion, it would seem that in a train moving with a speed of 240,000 kilometers per second, light would move five times more slowly in one direction than in a stationary train and 1.8 times faster in the other.

By studying the propagation of light, we ought to be able to find the absolute speed of the train.

Here is a ray of hope: Might it be possible to use the properties of light to define the notion of absolute rest?

A laboratory in which light spreads in all directions equally at a speed of 300,000 kilometers per second might be said to be at absolute rest. In any other laboratory moving in a straight line and uniformly relative to the first, the speed of light ought to be different in different

"ABSOLUTE REST"

directions. If this is so, nothing would remain of the relativity of motion, relativity of speeds, and relativity of rest established earlier.

"World Ether"

How are we to make sense of this state of affairs? In the past, use was made of the similarity in the behavior of sound and of light, and physicists introduced a special medium, called the ether, in which light was supposed to spread in the same way as sound spreads in air. It was assumed that in moving through the ether, objects avoided carrying the ether with them, just as a thin wire cage moving through water avoids carrying the water with it.

If our train is at rest relative to the ether, light will spread in all directions with the same speed. Any motion of the train relative to the ether will immediately show itself in the fact that the speed of light is different in different directions.

However, on introducing an ether—a medium whose vibrations appear as light—we find ourselves facing a number of perplexing questions. In the first place, the hypothesis itself is obviously artificial, for we can study the properties of air not only by observing the way sound travels in it but also by a variety of other physical and chemical methods, while the ether, on the other hand, succeeds in escaping any kind of physical detection in a very puzzling way. The density and the pressure of air can be measured by quite primitive experiments. But all attempts to find out something about the density or the pressure of the ether prove absolutely fruitless. The result is a rather absurd situation.

Undoubtedly any phenomenon of nature can be "explained" by introducing a special fluid with the required properties. But a true theory of phenomena is more than just the enumeration of known facts in learned language, precisely because it has many more consequences than follow directly from the facts on which it was founded. For instance, the notion of atoms came into science (broadly speaking) in connection with questions of chemistry, but the concept of an atom then made it possible to explain and predict an enormous number of phenomena not related to chemistry.

As for the suggestion of an ether, we could justly compare it with the attempt by a savage to explain the working of a phonograph by saying that the mysterious box contained a special "spirit of the phonograph."

Naturally, "explanations" of this kind explain nothing at all.

Even before the ether was invented, physicists had sad experiences of the same kind. The phenomenon of burning was "explained" in its time by the properties of a special fluid, the "phlogiston," and heat phenomena by another fluid, the "caloric." It is worth mentioning that, like the ether, both these fluids were distinguished by complete elusiveness.

A Difficult Situation Arises

The important thing is that if the principle of relativity of motion does not hold good for light, this inevitably means that the principle is violated by all other objects.

Any material medium exerts a resistance against the motion of objects. Therefore, the displacement of objects in the ether should also be connected with friction. The motion of any object would be slowed down and would end in a state of rest. In fact, however, the Earth (from geological knowledge) has been revolving around the sun for many thousands of millions of years and there are no signs of its being slowed down by friction.

So in attempting to explain the curious behavior of light in a moving train by the presence of an ether, we have reached a dead end. The notion of an ether does not remove the contradiction between the violation of the principle of relativity by light and its observance by all other motions.

Experiment Must Decide

What is one to do with this contradiction? First, let us note the following facts.

The contradiction that we found between the behavior of light and the principle of relativity of motion was reached exclusively by argument. It is true that our arguments were extremely convincing. But if we confined ourselves to argument alone, we would be like certain ancient philosophers who attempted to discover the laws of nature within their own heads. If you do this, you cannot avoid the danger that the world you construct, for all its merits, may prove to be extremely unlike the real world.

The supreme arbiter of any physical theory is always the experiment. Therefore, we should not go on discussing how light should spread in a moving train but should turn instead to experiments that show how in fact it does spread in these conditions.

In setting up such experiments, we have the advantage that we live on an object that is definitely moving. In its revolution around the sun the Earth moves on anything but a straight line, and therefore it cannot always be stationary from the point of view of any laboratory at rest.

Even if we take our starting laboratory to be one relative to which the Earth is at rest in January, it will certainly be in motion in July, because the direction of its motion around the sun changes. If, therefore, we study the propagation of light on the Earth, we shall in fact be studying the propagation of light in a moving laboratory the speed of which is extremely respectable by our standards: 30 kilometers per second. (We can forget about the rotation of the Earth around its axis, which leads to speeds of only about half a kilometer per second.)

Now may we identify the terrestrial globe with the moving train that carried us into this dead end? We assumed the train to be moving in a straight line and uniformly, but the Earth moves in an orbit. We are, however, justified in identifying the two. During the tiny fraction of a second taken by light to pass through our laboratory apparatus, it is perfectly correct to assume that the Earth moves uniformly in a straight line. The error introduced in making this assumption is so insignificant as to be undetectable.

Since we may compare the train and the Earth, it is natural to expect light to behave on the Earth in the same strange way as in our train: It should travel with different speeds in different directions.

The Principle of Relativity Triumphs

In 1881 such an experiment was performed by A. A. Michelson, one of the greatest experimenters of the last century. He made high-precision measurements of the speed of light in different directions relative to the Earth. To detect the expected very small difference in speeds, Michelson had to use extremely refined experimental techniques, and in this he showed tremendous ingenuity. The accuracy of the experiment was so great that it would have been possible to detect much smaller differences in speeds than were expected.

The result of Michelson's experiment proved to be quite different from what the foregoing reasoning has suggested. Michelson showed that on the moving Earth light spreads with exactly the same speed in all directions. In this respect the behavior of light is the same as the behavior of a bullet—it does not depend on the motion of the laboratory, and its speed relative to the walls of the laboratory is the same in all directions.

Thus Michelson's experiment showed that, contrary to all our arguments, the behavior of light does not in the least contradict the principle of relativity of motion. On the contrary, there is full agreement. In other words, our argument on pages 22 and 23 has proved to be in error.

Out of the Frying Pan into the Fire

Thus experiment has liberated us from the serious contradiction between the laws of light propagation and the principle of the relativity of motion. The contradiction has proved to be only apparent and is evidently due to a flaw in our reasoning. But where is this flaw?

For nearly a quarter of a century, from 1881 to 1905, physicists throughout the world racked their brains trying to answer this question, but all the explanations proposed led to more and more contradictions between theory and experiment.

If a source of sound and an observer are moved in a thin wire cage, the observer will feel a strong wind. If you measure the speed of sound relative to the cage, it will be less in the direction of the motion than in the opposite direction. But if you put the source of sound in a railway carriage and measure the velocity of sound there with the doors and windows closed, you will find that the speed of sound is the same in all directions, because the air is carried with the train.

Going from sound to light, one might make the following suggestion to explain the results of Michelson's experiment. Suppose that in moving through space the Earth does not leave the ether behind in passing through it, as the thin wire cage does for sound. Assume rather that the ether is carried with the Earth so that the ether and the Earth move as one whole. Then the result of Michelson's experiment becomes easy to understand.

Unfortunately, this assumption stands in sharp contradiction to a large number of other experiments—for instance, experiments on the way light travels through a pipe carrying a flow of water. If it were right to assume that the ether takes part in the motion of material objects, then a measurement of the speed of light in the direction of the flow of water should give a speed equal to the speed of light in water at rest plus the speed of the water. Direct measurement, however, gives a smaller value than that given by such reasoning.

We have already mentioned the extremely strange state of affairs whereby objects moving through the ether do not experience any noticeable friction. Yet, if they not only pass through the ether, but even carry the ether along, friction should most certainly be significant!

Thus all attempts to get around the contradiction resulting from the surprising result of Michelson's experiment were entirely unsuccessful.

Let us sum up.

Michelson's experiment confirms the principle of relativity of motion not only for ordinary objects but also for light—in other words, for all natural phenomena.

As we saw earlier, the principle of the relativity of motion leads directly to the relativity of speeds. For different laboratories moving relative to one another, speeds must be different. On the other hand, the speed of light, 300,000 kilometers per second, proves to be the same in all laboratories. Therefore, this speed is not relative but absolute!

4

Time Proves To Be Relative

Is There in Fact a Contradiction?

At first sight it may seem that we are faced with a purely logical contradiction. The fact that the speed of light is the same in all directions confirms the principle of relativity, but at the same time the speed of light itself is absolute.

But remember the attitude of medieval man to the fact that the Earth is a sphere: To him this seemed to be in sharp contradiction to the existence of a force of gravity, because all objects would apparently have to fall "down" off the Earth. But we know that in fact there is no logical contradiction at all. The notions of up and down are simply not absolute but relative.

We have exactly the same situation with the behavior of light.

It would be futile to look for a logical contradiction between the principle of the relativity of motion and the absolute nature of the speed of light. We get this contradiction only because we have unwittingly introduced some further assumptions, as medieval man did when he denied that the Earth is round and assumed up and down to be absolute. His belief in an absolute up and an absolute down seems ridiculous to us. It arose because in those days experimental possibilities were very limited. People traveled very little and knew only a small part of the Earth's surface. Evidently something similar has happened to us; because of our limited experience, we seem to have taken something to be absolute which is in fact relative.

What is this something?

To discover our mistake we shall from now on rely only on statements that can be checked by experiment.

We Go on a Train Journey

Let us imagine a train 5,400,000 kilometers long moving in a straight line with a uniform speed of 240,000 kilometers per second. Assume that at a certain instant a light is switched on in the middle of the train. The first and last carriages have automatic doors which open as soon as light falls on them. What are people on the train going to see and what will people on the platform see?

As agreed, we shall rely only on experimental facts in answering this question.

People sitting in the middle of the train will see the

following: Since, according to Michelson's experiment, light travels with the same speed in all directions relative to the train—namely, at 300,000 kilometers per second—the light will reach the first and last carriages simultaneously, in nine seconds (2,700,000 ÷ 300,000). Therefore, both doors will open together.

But what will people on the platform see? Relative to the station, light also travels with a speed of 300,000 kilometers per second. But the rear carriage is moving toward the beam of light. Therefore, the light will arrive at the rear carriage in

$$2,700,000 \div (300,000 + 240,000) = 5 \text{ seconds.}$$

As for the first carriage, the light must catch up with it and therefore will reach it only after

$$2,700,000 \div (300,000 - 240,000) = 45 \text{ seconds.}$$

Thus it will appear to people on the platform that the doors on the train do not open simultaneously. The door at the rear will open first and the door at the front only after $45 - 5 = 40$ seconds.*

Thus we see that two completely similar events, the opening of the front and back doors of the train, will be simultaneous for people in the train and 40 seconds apart for people on the platform.

Common Sense in Disgrace

Is there a contradiction in this? Surely the state of affairs we have described is completely absurd—like saying

---

* Later on these arguments are put forward somewhat more precisely; (see p. 55).

that the length of a crocodile from tail to head is two meters and from head to tail one meter!

Let us try to get quite clear why our result seems so absurd to us even though it agrees entirely with experimental facts.

No matter how much we argue, we shall not succeed in finding a logical contradiction in the conclusion that two phenomena which happen at the same time for people in the train are separated by an interval of 40 seconds for people on the platform. The only thing we can tell ourselves in consolation is that our deductions defy "common sense."

But remember how the "common sense" of medieval man resisted accepting the fact that the Earth revolves around the sun. All his daily experience did, with the greatest assurance, persuade medieval man that the Earth is at rest, and that the sun moves around it. And surely it was also common sense that prompted that ridiculous proof, mentioned above, which was intended to show that the Earth could not possibly be round.

The clash of "common sense" with actual facts is illustrated amusingly by the well-known anecdote of the rustic who sees a giraffe at the zoo and exclaims, "There can't be such a thing."

So-called common sense represents nothing but a simple generalization of the notions and habits that have grown up in our daily life. It is a definite level of understanding reflecting a particular level of experiment.

The whole difficulty in grasping and accepting the fact that for people on the platform two events will appear non-simultaneous although they happen simultaneously on the train is like the difficulty of the rustic puzzled by the sight of a giraffe. Like the rustic who never saw such an

animal, we have never moved with a speed of 240,000 kilometers per second. There is nothing surprising in the fact
that when physicists meet such fabulous velocities, they
observe things very different from those we are used to in
daily life.

The unexpected result of Michelson's experiment
made physicists face these new facts and forced them, de-
spite common sense, to re-examine such seemingly obvious
and familiar notions as that of two events happening at
the same time.

Of course you could stick to your basis of "common
sense" and deny the existence of the new events, but if you
did, you would be behaving like the rustic in the story.

Time Suffers the Same Fate as Space

Science is not afraid of clashes with so-called common sense. It is only afraid of disagreement between existing ideas and new experimental facts, and when such disagreement occurs, science relentlessly smashes the ideas it has previously built up and raises our knowledge to a higher level.

We assumed before that two simultaneous events are simultaneous in any two laboratories. Experiment has led us to another conclusion. It has become clear that this is true only in the event of the two laboratories being at rest relative to each other. If, however, two laboratories are in motion relative to each other, events that are simultaneous in one of them must be assumed to be non-simultaneous in the other. The notion of simultaneous events becomes relative; it has meaning only if we indicate the motion of the laboratory from which the events are observed.

Let us remember the example of the relativity of angular sizes discussed on page 5. What was the situation there? Assume that the angular distance between two stars observed from the Earth turns out to be zero because the two stars lie on the same line of vision. In daily life we shall never be led to a contradiction if we assume that this statement is an absolute. But things become different if we leave the limits of the solar system and observe the same two stars from some other point in space. The angular distance will then turn out to be different from zero.

To modern man it is obvious that two stars which coincide when observed from the Earth may not coincide if the observation is made from other points in space, but this would have seemed absurd to a medieval man who pictured the sky as a star-studded dome.

Assume you are asked: But how is it in reality—forgetting about laboratories? Are the two events simultaneous or aren't they? Unfortunately this question has no more sense than the question: Do the two stars in reality —forgetting about points of observation—lie on the same line of vision or not? The crucial fact is that finding the two stars on the same line is a matter not only of their positions but also of the point from which they are observed. In the same way the simultaneous occurrence of two events is a matter depending not only on the events but also on the laboratory in which the events are observed.

Up to now we have been dealing with speeds which are small compared to the speed of light, and because of this it was impossible to show that the notion of simultaneous events is a relative concept. Only when we begin to study motions at speeds comparable to the speed of light are we forced to reconsider the concept of simultaneous events.

In the same way, people were obliged to reconsider the notions of up and down when they began to travel over distances comparable to the size of the Earth.

True, we still have no means of moving with speeds close to the speed of light, so that in our personal experience we cannot observe the happenings just described, which are so paradoxical from the point of view of our old notions. But thanks to modern experimental techniques we can demonstrate these facts with certainty in many physical phenomena.

Thus the fate that befell space has now overtaken time! The words "at one and the same time" have turned out to be as meaningless as the words "at one and the same place." If you want to state the time interval between two events, you must specify the laboratory with respect to which the statement is made, just as is required in stating the distance in space between the two events.

Science Triumphs

The discovery that time is relative brought about a profound revolution in man's picture of nature. It represents one of the greatest victories of the human mind over the distorted notions acquired over the ages. It can be compared only with the revolution in human ideas brought about by the discovery of the fact that the Earth is a sphere.

The discovery that time is relative was made in 1905 by the greatest physicist of the twentieth century, Albert Einstein (1879 to 1955). This discovery placed the 26-year-old Einstein among the Titans of human thought. We remember him as the equal of Copernicus and Isaac Newton, those other pioneers of new paths in science. V. I. Lenin called Albert Einstein one of the "greatest transformers of our knowledge of nature."

The science of the relativity of time and the consequences following from it is usually called the Theory of Relativity. It should not be confused with the principle of the relativity of motion.

Speed Has a Limit

Before the Second World War, airplanes flew with speeds less than the speed of sound; nowadays we build "supersonic" aircraft. Radio waves travel with the speed of light. Could we not try to create a super-telegraphy in which signals are transmitted with speeds greater than the speed of light? This proves to be impossible.

For if it were possible to transmit signals with infinite

speed, we would find a means of establishing uniquely that two events are simultaneous. We could say that two events are simultaneous if an infinitely fast signal marking the first event arrived simultaneously with a signal marking the second. In this way the property of occurring simultaneouly would acquire an absolute character, independent of the motion of any laboratory in which the statement was made.

But since the absolute nature of time is denied by experiment, we conclude that the transfer of signals cannot be instantaneous. The speed at which an action can be transmitted from one point in space to another cannot be infinite, which means that it cannot exceed a certain finite quantity called the limiting speed.

This limiting speed is the same as the speed of light.

For, according to the principle of the relativity of motion, the laws of nature must be the same in all laboratories that move uniformly and in a straight line relative to one another. The statement that no speed can exceed the given limit is also a law of nature, and therefore the value of the limiting speed must be exactly the same in different laboratories. As we know, the speed of light has just this property.

Thus the speed of light is not simply the speed of travel of a certain natural phenomenon. It plays the very important role of a limiting speed.

The discovery of the existence of a limiting speed in the world is one of the greatest triumphs of the human mind and of the experimental capabilities of man. A physicist of the last century could not have made this discovery, nor could he have concluded that the existence of this limiting speed in the world can be proved. What is more, even if in his experiments he had hit upon the existence of a limiting speed in nature, he could not have been sure that this was a law of nature rather than merely a result of limitations of his experimental methods, which might be improved with the development of his technique.

The principle of relativity shows that the existence of a limiting speed is intrinsic in the very nature of things. To expect that the progress of technology will enable us to reach speeds exceeding the speed of light is as ridiculous as thinking that the absence on the Earth of points farther apart than 20,000 kilometers is not a law of geography but a limitation of our knowledge and to hope that with the development of geography we might succeed in finding places farther away from each other.

The speed of light plays such an exceptional part in nature just because this speed is the limit for the movement of anything whatsoever. Light either outpaces any other phenomenon or, in the extreme case, is equaled by it.

If the sun were to split in two and form a double star, the motion of the Earth would certainly change. A physicist of the last century, not knowing of the existence of a limiting speed in nature, would certainly assume that a change in the motion of the Earth would take place instantaneously after the sun had split. But light would require eight minutes to arrive on the Earth from the broken sun. In reality, then, changes in the motion of the Earth would also begin only eight minutes after the sun had split, and until that moment the Earth would move just as if the sun had remained intact. Quite generally, any event happening to the sun or on the sun cannot have any effect on the Earth or on its motion until these eight minutes have elapsed.

The finite speed with which signals travel does not, of course, rob us of the possibility of establishing that two events are simultaneous. We simply have to take into account the time of delay of the signal, as is done quite commonly.

However, such a method for establishing that two events are simultaneous is now completely compatible with the relative nature of this notion. For in order to calculate the delay time, we must divide the distance between the places at which the events took place by the speed of travel of the signal. And we have seen, when discussing the question of sending letters from the Moscow-Vladivostok Express, that even the position in space is a completely relative notion!

Earlier and Later

Let us assume that in our train with its flashing lights (which we shall call Einstein's train) the mechanism of the automatic doors goes wrong and the people in the train notice that the forward door opens 15 seconds before the rear door. The people on the station platform, on the other hand, will see the rear door opening $40 - 15 = 25$ seconds earlier. So everything that took place earlier for one laboratory may take place later for the other.

However, it strikes one immediately that such relativity in the notions "earlier" and "later" must have limits. For instance, we could hardly accept—from the point of view of any laboratory whatsoever—that a child is born before his mother.

A sunspot appears. An astronomer observing the sun through his telescope sees the spot eight minutes later. Anything the astronomer does after this will be absolutely later than the appearance of the spot—later from the point of view of any laboratory from which both the sunspot and the astronomer are observed. Conversely, everything that happened to the astronomer at times earlier than eight minutes before the appearance of the spot happened absolutely earlier.

But if the astronomer, shall we say, put on his spectacles at a moment between these two limits, the time relation between the appearance of the spot and his putting on his spectacles is not absolute. We may be moving relative to the astronomer and the sunspot in such a way that we see the astronomer putting on his spectacles before the spot appears, after it appears, or at the same time, according to the speed and direction of our motion.

Thus the principle of relativity shows that there are three types of time relation between events: The absolutely earlier, the absolutely later, and the "neither earlier nor later"—or more precisely, the earlier or later depending on what laboratory the events are observed from.

5

Clocks and Rulers Play Tricks

We Get on the Train Again

Ahead of us is a very long railway line with Einstein's train moving along it. At a distance of 864,000,000 kilometers from each other there are two stations. At its speed of 240,000 kilometers per second, Einstein's train needs an hour to cover this distance.

There is a clock at each of these stations. A passenger boards the train at the first station and before its departure sets his watch by the station clock. On arriving at the second station, he notices with astonishment that his watch is slow.

The watchmaker had assured the passenger that his watch was in perfect order.

What has been going on?

To sort this out, let us imagine that the passenger directs a beam from a flashlight fixed to the floor of the carriage onto the ceiling. There is a mirror on the ceiling from which the beam of light is reflected back to the flashlight. The path of the ray of light, as seen by the passenger in the carriage, is drawn in the upper half of the figure on page 45. This path looks quite different for the observer on the platform. While the ray passes from the flashlight to the mirror, the mirror moves because the train moves. While the ray is returning, the flashlight covers the same distance again.

We see that for the observer on the platform, the ray of light traveled a greater distance than for the observer in the train. We know on the other hand that the speed of light is an absolute speed, the same both for travelers on the train and for people on the platform. This forces us to the conclusion that at the station more time elapsed between the departure and the return of the ray of light than in the train!

It is simple to calculate the ratio of these times.

Let us assume that the observer on the platform established that ten seconds elapsed between the departure and the return of the ray. During these ten seconds the ray traversed $300,000 \times 10 = 3,000,000$ kilometers. It follows that the sides AB and BC of the isosceles triangle ABC (in the lower drawing on page 45) are each 1,500,000 kilometers long. The side AC is obviously equal to the distance traveled by the train in ten seconds, which is $240,000 \times 10 = 2,400,000$ kilometers.

Now it is easy to determine the height of the carriage; it is just the height BD of the triangle ABC.

Remember that in a right-angled triangle (ABD), the square of the hypotenuse (AB) is equal to the sum of the

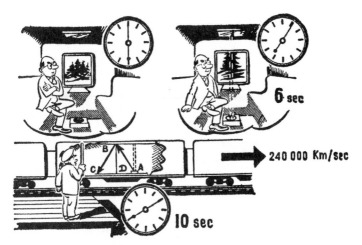

squares of the other two sides (AD and BD). From the
equation $AB^2 = AD^2 + BD^2$ we find that the height of
the carriage is:

$$BD = \sqrt{AB^2 - AD^2}$$
$$= \sqrt{1,500,000^2 - 1,200,000^2}$$
$$= 900,000 \text{ kilometers.}$$

Quite a size, but not surprising in view of the astronomic
dimensions of Einstein's train.

The path traveled by the ray from the floor to the ceiling and back, from the point of view of the passenger, is
evidently equal to twice this height, *i.e.*, to $2 \times 900,000$
$= 1,800,000$ kilometers. To travel this distance light needs
$1,800,000 \div 300,000 = 6$ seconds.

Clocks Go Slow Systematically

Thus, while ten seconds elapsed at the station, only
six seconds went by in the train. If according to station

time the train arrived an hour after its departure, the time elapsed according to the passenger's watch is only 60× 6/10 = 36 minutes. Thus in one hour the traveler's watch lost 24 minutes compared to the station clock.

It is not difficult to guess that the slowing down of the watch will increase as the train speeds up. The closer the speed of the train to the speed of light, the closer the length of the side AD, representing the path of the train, will be to the hypotenuse AB, representing the path traversed by the ray of light during the same time. Correspondingly, the ratio of the side BD to the hypotenuse diminishes. But this ratio is just the ratio of the times in the train and at the station. As we make the speed of the train approach the speed of light, we can make the time elapsing in the train during an hour of station time as small as we like. If the speed of the train is 0.9999 of the speed of light, only one minute will elapse in the train in an hour of station time!

So any clock in motion will run slow compared to a clock at rest. But doesn't this result contradict the principle of the relativity of motion, which was our starting point? Doesn't this mean that the clock that goes faster than any other is in a state of absolute rest?

No, because we compared the watch in the train with the clocks at the stations in completely unequal conditions. We used not two but three timepieces! The traveler compared his watch with two different clocks at two different stations. And, conversely, if there had been two clocks, one at the front and one at the rear of the train, an observer at one of the stations, comparing the indications of the station clock with the readings of the clocks seen through the windows of the passing train, would discover that the station clock was running slow systematically.

For in this case—given uniform straight-line motion of the train relative to the station—we are entitled to consider the train to be at rest and the station to be moving. The laws of nature must be the same in both cases.

Any observer at rest relative to his own timepiece will see that other clocks moving with respect to him run fast—the greater their speed, the faster they are.

This statement is quite similar to that by two observers standing near telegraph poles, each of whom will say that his pole is seen at a greater angle than the other fellow's.

The Time Machine

Let us now assume that Einstein's train moves not along a trunk line but on a circular railway, and that after a certain time it returns to its place of departure. As we have already seen, the passenger in the train will discover that his watch goes slow—the faster the train moves, the slower his watch gets. By increasing the speed of Einstein's train on the circular railway, we can reach a situation in which while no more than an hour elapses for the passenger, many years will pass for the stationmaster. Returning to his place of departure after one day (according to his own watch!), our passenger will find that all his friends and relations have long since died!

In contrast to the case of travel between two stations, in which the passenger checks his watch by two different clocks, here on the circular journey the readings of only two clocks, not three, are compared—namely, the clock in the train and the clock at the station from which the journey started.

Is this not in contradiction with the principle of relativity? Can we assume that the passenger is at rest and the station takes a round trip with the speed of Einstein's train? If we could, we would reach the conclusion that the people at the station would see only one day pass while the man on the train would experience many years. But this argument would be incorrect, and this is why.

We explained earlier that only objects not acted on by any force can be considered to be at rest. True, there is not just one but an innumerable manifold of "states of rest," and as we saw, two bodies at rest may move uniformly and in a straight line relative to each other. But the watch in Einstein's train moving on a circular railway is most certainly acted on by centrifugal force, so we must definitely not assume that it is at rest. In this case there is an absolute difference between the indication of the station clock, which is at rest, and the watch in Einstein's train.

If two people with watches indicating the same time

separate and meet again after a certain time, the longer time will be indicated by the watch of the person who was at rest or who moved uniformly and in a straight line—that is, by the watch which did not experience any forces.

A journey on a circular railway with a speed close to the speed of light allows us in principle to make H. G. Wells' "time machine" come true in a limited sense. Disembarking at our place of departure, we find that we have moved into the future. True, with this time machine we can transport ourselves into the future but are unable to return to the past. This is its great difference compared to Wells' machine.

It is futile even to hope that future developments of science will allow us to travel into the past. Otherwise we would have to accept the possibility in principle of quite absurd situations. Traveling into the past, we could, for example, find ourselves in the position of being people whose parents have not yet seen the light of day!

On the other hand, a journey into the future involves only apparent contradictions.

Journey to a Star

There are stars in the sky whose distance from us is such that a ray of light from one of these stars takes, say, 40 years to reach the Earth. Since we know already that it is impossible to move with speeds greater than the speed of light, we could conclude that it would be impossible to get to such a star in less than 40 years. But this deduction is wrong, because it does not take into account the change in time connected with the motion.

Assume that we are flying toward the star in an Einstein rocket with a speed of 240,000 kilometers per second. To someone on the Earth we shall reach the star in 300,000 × 40 ÷ 240,000 = 50 years.

But if we are flying in Einstein's rocket at the speed assumed, this time will be shortened in the ratio of 10 to 6. Therefore we shall reach the star not in 50 years but in 6/10 × 50 = 30 years.

By increasing the speed of the Einstein rocket and making it approach the speed of light, we can shorten by as much as we like the time needed by the travelers to reach a star at that distance. Theoretically, by flying sufficiently fast, we could reach the star and return to Earth in one minute! On the Earth, however, 80 years will elapse, whatever we do.

It may seem that this opens up the possibility of prolonging human life, even if only from the point of view of other people (since a man ages according to his "own" time). Unfortunately, closer examination proves that the outlook is very far from promising.

In the first place, the human organism is not adapted to tolerating, for any long period, acceleration much higher than the acceleration of gravity on the Earth. To accelerate to a speed approaching the speed of light at this rate would take a very long time. Calculations show that on a voyage lasting six months with an acceleration equal to gravity, you would gain only six weeks. If you make the voyage longer, your gain in time will increase rapidly. By traveling in a rocket for a year, you would gain another year and a half, and a voyage lasting two years would gain you 28 years. If you stay on a rocket for three years, more than 360 years will elapse on the Earth!

The figures look encouraging.

But things are not so good in the matter of energy consumption. With a one-ton rocket, a most modest weight, the energy required to fly at a speed of 260,000 kilometers per second (the speed needed to "double" time, so that for each year of travel in the rocket, two years elapse on the Earth) is 250,000,000,000,000 kilowatt-hours. This is the total amount of energy generated on the whole of the Earth during several months.

But we have only worked out the energy of the rocket in flight. We have not taken into account that we must first bring our flying machine up to the speed of 260,000 kilometers per second! And at the end of the voyage the rocket must be slowed down for safe landing. How much energy is needed for this?

Even if we had a fuel that could produce an engine jet of the greatest possible speed—the speed of light—the energy would have to be 200 times the amount calculated above. We would have to spend as much energy as is produced by all mankind during several decades. The speeds of actual rocket jets are tens of thousands of times smaller than the speed of light. This makes the energy expenditure needed for our imaginary flight fantastically large.

Objects Get Shortened

So we see that time has come off the pedestal of an absolute concept and has only a relative meaning, requiring a precise indication of the laboratory in which measurements are made.

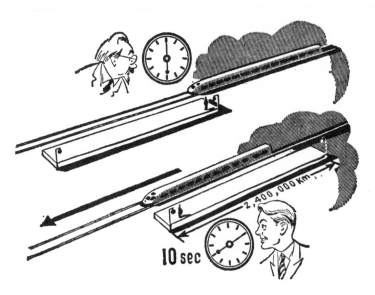

10 sec

Let us now turn to space. Even before we described Michelson's experiment, we agreed that space is relative, but in spite of this we still ascribed absolute character to the dimensions of objects. In other words, we assumed that dimensions are attributes of an object and are independent of what laboratory is used for observation. But the theory of relativity forces us to say goodbye to this view. Like the notion of absolute time, it is a mere prejudice which grew up because we have always had to do with speeds negligibly small compared to the speed of light.

Imagine that the Einstein train is passing a station platform 2,400,000 kilometers long.

Will the passengers in the Einstein train agree with this statement? According to the reading of the station clock, the train will pass from one end of this platform to the other in 2,400,000 ÷ 240,000 = 10 seconds. But the passengers have their own watches, according to which the motion of the train from one end of the platform to the other will take less time. As we know already, it will take six seconds. The passengers will have every right to conclude from this that the length of the platform is not 2,400,000 kilometers at all but 240,000 × 6 = 1,440,000 kilometers.

We see that the length of the platform is greater from the point of view of a laboratory at rest relative to it than for a laboratory relative to which it is moving. Any moving object will be shortened in the direction of its motion.

However, this shortening is by no means a sign of

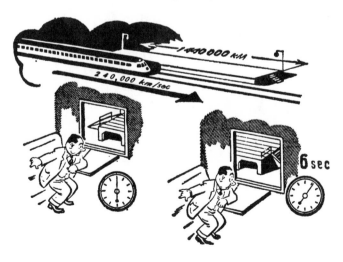

absolute motion; we have only to place ourselves in a laboratory at rest relative to the object to make it become longer again. In the same way, the passengers will find the platform shortened, while to people on the platform the Einstein train will appear shortened (in the ratio 6 to 10).

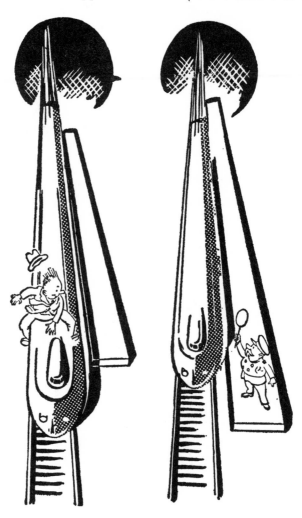

And this will be no optical illusion. The same thing would show up with any device with which we chose to measure the length of an object.

In connection with this shortening of objects, we must now introduce a correction to our discussion on page 33 concerning the time at which the doors on the Einstein train open. When we calculated the instant at which the doors opened, from the point of view of the observer on the station platform, we assumed that the length of the moving train would be the same as for a train at rest. In fact, however, the train was shorter for the people on the platform. Correspondingly, from the point of view of the station clock the time interval between the opening of the doors will actually be not 40 seconds but only 6/10 × 40 = 24 seconds. Of course this correction in no way invalidates our previous results.

The drawings on the opposite page show the Einstein train and the station platform as seen by observers on the train and on the platform. The drawing on the right shows the platform longer than the train, while on the left the train is longer than the platform. Which of these pictures corresponds to reality?

The question is as devoid of sense as the question on page 4 about the shepherd and the cow. Both pictures are of one and the same object, "photographed" from different points of view.

Velocities Play Tricks

What is the speed of a passenger relative to the railway track if he is walking toward the front of the train at a speed of five kilometers per hour while the train is moving at 50 kilometers per hour? Surely the speed of the passenger relative to the track is 50 + 5 = 55 kilometers per hour. The reasoning used here is based on the law of addition of speeds, and we do not doubt the truth of this law. For in an hour the train will cover 50 kilometers and the man in the train another 5 kilometers. This gives us the 55 kilometers we were just talking about.

But it is quite clear that the existence of a speed limit (the speed of light) robs this law of universal applicability to all speeds, large and small. For if the passenger were moving in the Einstein train at a speed of, say, 100,000 kilometers per second, his speed relative to the railway track could not be equal to 240,000 + 100,000 = 340,000 kilometers per second, because this would be more than the limiting speed of light and so would not exist in nature.

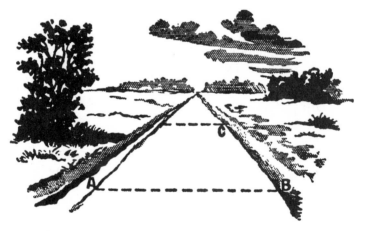

Thus the law of addition of speeds used in daily life proves to be inaccurate. It holds only for speeds that are sufficiently small compared with the speed of light.

The reader by now will be prepared for all sorts of paradoxes in the theory of relativity and will easily understand the reason why the argument just used in adding speeds is not valid. To apply it, we had to add a distance covered in an hour by the train along the track and one covered by the passenger in the train. The theory of relativity shows that we cannot add these distances. To do so would be just as absurd as if we tried to determine the area of the field shown in the picture on page 56 by multiplying the lengths of the sides AB and BC and forgetting that the latter is distorted by perspective. Also, if we want to determine the speed of the passenger relative to the station, we must determine the path he has traversed in an hour according to station time, while for measuring the speed of the passenger in the train we have used train time. We know already that the two are by no means the same.

All this leads to the result that speeds of which at least one is comparable to the speed of light must be added differently from the way speeds are usually added. This paradox of the addition of speeds can be shown in experiment, for instance, by observing how light travels in moving water (a question mentioned earlier). The fact that the speed with which light travels in moving water is not equal to the sum of the speed of light in water at rest and the speed of motion of the water, but is less than the sum, is a direct consequence of the theory of relativity.

Speeds add together in a specially peculiar way in the case when one of them is exactly equal to 300,000 kilometers per second. As we know, this speed has the property of remaining unchanged, whatever the motion of the laboratory in which we observe it. In other words, no matter what speed we add to the speed of 300,000 kilometers per second, we get the same speed—300,000 kilometers per second.

A simple analogy illustrates the fact that the usual law of addition of speeds is not valid.

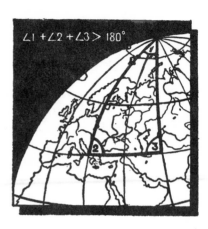

$\angle 1 + \angle 2 + \angle 3 > 180°$

$\angle 1 + \angle 2 + \angle 3 = 180°$

It is well known that in a plane triangle (see the drawing above) the sum of the angles 1, 2, and 3 equals two right angles. But imagine a triangle drawn on the surface of the Earth (the figure on the right). Because the Earth is a sphere, the sum of the angles of such a triangle will be greater than two right angles. This is true for any triangle drawn on a spherical surface, but the difference becomes noticeable only when the sphere is very large, say comparable to the dimensions of the Earth.

Just as one can use the laws of ordinary plane geometry in measuring areas of small regions on the Earth, so one can use the ordinary law of addition of speeds if the speeds in question are small.

6

Work Changes Mass

Mass

Assume that we have an object at rest and want to make it move with a definite speed. To do this we must apply a force to it. If the resulting motion is not opposed by extraneous forces, such as friction, the object set in motion will move with ever-increasing speed. The time needed to accelerate the object to the speed we want, if we use a certain given force, will depend on the mass or weight of the object.

Let us imagine that we have, in frictionless outer space, two balls of equal size—one of lead, the other of wood. Let us pull these balls with the same force until each attains a speed of, say, ten kilometers per second. Clearly, to get this result we shall have to let the force act on the lead ball a longer time than on the wooden ball. To characterize this state of affairs we say that the lead ball has a greater mass than the wooden ball. When a constant force is applied, speed increases in proportion with time; therefore we can use the ratio of the time required to reach a given speed to the speed itself to measure mass. The mass of an object is proportional to this ratio, the coefficient of proportionality depending on the forces producing the motion.

Mass Increases

One of the most important properties of any object is its mass. We are used to assuming that the mass of a body does not change. In particular, it does not depend on speed. This follows from our first statement that when a constant force acts, the speed increases in proportion to the time during which the force acts. This statement is based on the usual law of addition of speeds. But we have just proved that this law is not valid in all conditions.

What do we do to obtain the value of the speed at the end of the second second, while the force acts? We add the speed that the body had at the end of the first second to the speed it acquired during the second second—according to the usual law of addition of speeds.

But we can proceed in this way only up to the point where the speeds acquired become comparable with the speed of light. When this happens we can no longer use the old rules. If we add the speeds according to the theory of relativity, we get an answer that is always a little smaller than we would have found from the old incorrect law of addition. This means that if the speed reached is high, it will not increase in proportion with the time of the force's action, but more slowly. This is clear from the fact that there is a limiting speed.

As the speed of an object approaches the speed of light, under the action of a constant force, it increases more and more slowly, so that the limiting speed is never exceeded.

As long as we could assume that the speed of an object increases in proportion to the time of action of a force, mass could be assumed to be independent of speed. But when the speed of an object becomes comparable to the speed of light, the proportionality between time and speed breaks down, so that mass begins to depend on speed. Because the time of acceleration can increase without limit, but the speed cannot exceed the limiting speed, we see that the mass must increase with speed. And the mass becomes infinite when the speed of the object equals the speed of light.

Calculation shows that in any motion the mass of the object increases by as much as its length decreases. Thus the mass of the Einstein train moving at 240,000 kilometers per second is 10/6 times greater than the mass of the train at rest.

Naturally, in dealing with ordinary speeds (small compared to the speed of light) we can neglect changes in mass completely, just as we neglect the dependence of the size of the body on its speed or the dependence of the time interval between two events on the speed with which the person observing the events moves.

The dependence of mass on speed given by the theory of relativity can be verified directly by studying the motion of fast electrons. Under modern experimental conditions, electrons moving with a speed near the speed of light are not a rarity but a daily occurrence. In special accelerators, electrons can be pushed up to speeds that differ from the speed of light by less than 30 kilometers per second. Thus modern physics is able to compare the mass of an electron moving at enormous speed with the mass of an electron at rest. And the results of experiments have completely confirmed the dependence of mass on speed given by the principle of relativity. At such very high speeds the electron's increase in mass with speed has definitely been detected and measured.

The increase in the mass of an object is closely connected with the work performed on the object: It is proportional to the work required in order to set the object in motion. It is not essential that the work be spent only on setting the body in motion. Any work performed on a body, any increase of the energy of the body, increases its mass. For instance, a hot body has a greater mass than a cold one, a compressed spring a greater mass than a free spring. It is true that the coefficient of proportionality relating the change in mass to the change in energy is very small: To increase the mass of an object by one gram (about 1/28 of an ounce), one must give it an energy of 25,000,000 kilowatt-hours.

This is why under ordinary conditions the change of mass of objects is extremely insignificant and escapes even the most accurate measurements. For instance, if a ton of water is heated from freezing to boiling, this leads to an increase of its mass of only about five-millionths of a gram.

If you burn a ton of coal in a closed furnace, the combustion products after cooling will have a mass only 1/3000 of a gram less than that of the carbon and oxygen from which they were formed. This missing mass was turned into the heat generated in burning the coal.

In this case the conversion of mass to energy is negligible. However, modern physics also knows cases in which the change in the mass of objects is very significant. This is so when atomic nuclei collide and new nuclei are formed from the colliding ones. For instance, when the nucleus of a lithium atom collides with the nucleus of a hydrogen atom and they form two helium atoms, the total mass is 1/400 less than before.

We have said that to increase the mass of an object by one gram, an energy of 25,000,000 kilowatt-hours must be put into it. Hence in the transmutation of one gram of a mixture of lithium and hydrogen into helium the amount of energy generated is: $25,000,000 \div 400 = 60,000$ kilowatt-hours!

Summing Up

We have seen that rigorous and convincing experiments force us to accept the truth of the theory of relativity, which reveals some of the amazing properties of the world around us—properties that escape us in initial (or, to be more precise, superficial) study.

We have seen what deep, radical changes the theory of relativity has wrought in some of man's basic notions and ideas. Does this not signify the complete collapse of the old, familiar ideas? Does it mean that the whole of the physics in existence before the emergence of the principle of relativity should be scrapped and thrown away, like an old boot that has served its useful time but is no longer of use to anybody?

If that were so, it would be futile to pursue scientific investigation. We would never be sure that in the future a new doctrine might not turn up, completely overthrowing everything that had gone before.

Let us imagine a passenger traveling in an ordinary train who decided to introduce corrections for relativity, fearing that otherwise his watch would be slow compared to the station clock. We would laugh at such a passenger. For the realistic correction would amount to only an infinitesimal part of a second—far less than the effect of a single jerk in the motion of the train upon his watch.

A chemical engineer who doubted whether the quantity of water he was heating retained a constant mass would need to have his head examined. But a physicist who failed to take into account the change of mass in nuclear transmutations would be sacked for incompetence.

Engineers will continue to plan their machines on the basis of the laws of classical physics, because the corrections for the theory of relativity have a much smaller influence on their engines than a single microbe settling on a flywheel. But a physicist observing fast electrons must certainly take into account the change of the mass of the electrons as they accelerate.

Thus the theory of relativity does not contradict but only deepens the ideas and concepts created in older science. It determines the limits within which these older ideas can be applied without leading to incorrect results. None of the laws of nature discovered by physicists before the birth of the theory of relativity are changed, but the limits of their application are now clearly marked out.

The relationship between the physics that takes into account the theory of relativity—which is called relativistic physics—and the older physics—which is called classical—is comparable to that between higher geodesy, which takes into account the spherical shape of the Earth, and lower geodesy, which neglects the spherical shape. Higher geodesy must start out from the relativity of the concept of the vertical; relativistic physics must take into account the relativity of the dimensions of a body and of time intervals between two events. Just as higher geodesy is a development of lower geodesy, so is relativistic physics a development and widening of classical physics.

We can perform the transition from the formulae of spherical geometry—the geometry on the surface of a sphere—to the formulae of plane geometry if we assume that the radius of the Earth is infinitely large. Then the Earth is no longer taken to be a sphere but an infinite plane; the vertical assumes an absolute meaning, and the sum of the angles of a triangle proves to be exactly two right angles.

Similarly, we can make a transition from relativistic to classical physics by assuming that the speed of light is infinitely large—*i.e.*, that light spreads instantaneously. If light spreads instantaneously, then, as we have seen, the notion of simultaneous events becomes absolute. Time intervals between events and sizes of objects acquire an absolute meaning, without reference to the laboratories from which they are observed. In this way all classical concepts can be preserved. And in our ordinary affairs, we can take the speed of light to be infinite, for all practical purposes.

However, any attempt to reconcile the actually finite speed of light with the old notions of space and time would put us in the position of the person who knows that the Earth is a sphere but is certain that the vertical in his native city is the absolute vertical and who therefore is afraid of moving far from his home for fear of falling head over heels into empty space.

Index

Index

A

absolute rest, concept of, 23-24
absolutes and relatives, 7
acceleration and deceleration, 17
addition of speeds, law of, 55-58
air, transmission of sound waves through, 22
angular dimensions, and size, 4-7, 36
Aristotle, 19
astronomy, and angular dimensions, 5-7
atomic nuclei, and mass, 62

B

bigger and smaller, meaning of, 4-5

C

"caloric," fluid, 26
Catholic Church; see Roman Catholic Church
clocks, and relativity, 43-58
"common sense," 8-9, 33-35
Copernicus, 38

D

day and night, meaning of, 4
deceleration and acceleration, 17
definitions:
  bigger and smaller, 4-5
  day and night, 4
  direction, 3-4
  earlier and later, 41-42
  rest, 14
  right and left, 3-4
  up and down, 7, 31-32
direction, meaning of, 3-4
displacement in space, 11-12
distances, measurements of, 43-58
down and up, meaning of, 7, 31-32

E

earlier and later, concepts of, 41-42
Earth:
  and meaning of up and down, 7, 31-32
  orbit speed, 20
  use of orbit to measure distances, 7